European Medical Device Regulation (MDR) for MedTech and Medical Device Manufacturers

I0391490

Des O'Brien

ISBN: 9781092813518

This guide does not purport to be an interpretation of law and/or regulations and is for guidance purposes only.

Regulations and the content herein are subject to change. In light of the repeal of medical device directives and the introduction of the new MDR requirements. It is solely the responsibility of the manufacturer to fulfil all regulatory, legal and product safety requirements.

Contents

Part 1

European MDR

Regulatory Framework and Context

The new European regulations on medical devices and in vitro medical devices were adopted on 05 April 2017 and came into force on 25th May 2017. Both these 2 new regulations replace and repeal Council Directives 90/385/EEC, 93/42/EEC Directive 98/79/EC and Commission Decision 2010/227/EU. Although adopted and in force, the new rules shall only apply after a 3-year transitional period, whereby regulations will enter into force in April 2020 for medical devices and for five years after entry into force (April 2022) for the Regulation on in-vitro diagnostic medical devices. The core goal of the new MDR rules and regulations is aimed at establishing a modern and robust EU legislative framework to ensure better patient safety and quality from manufacturers.

Improvements on Historical Directives

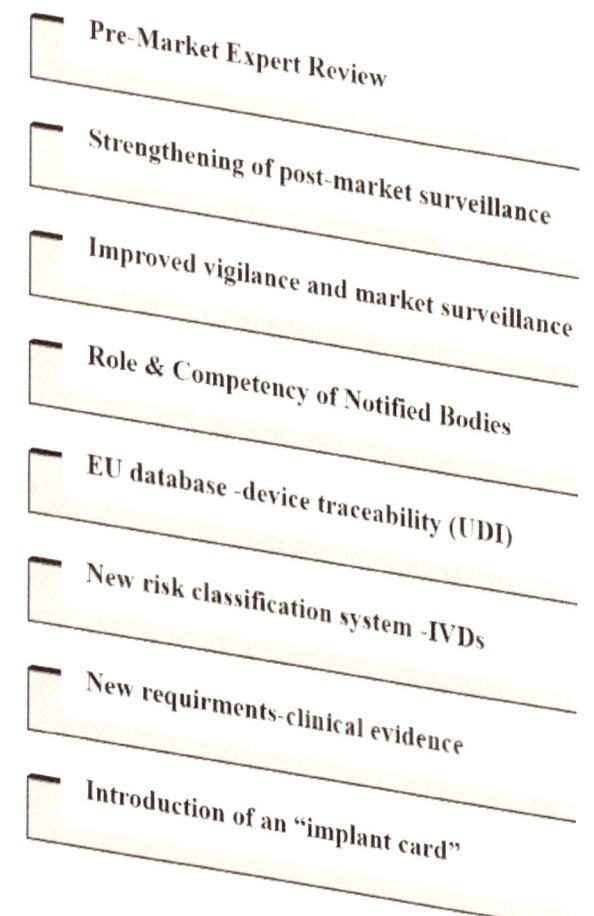

Pre-Market Expert Review

Strengthening of post-market surveillance

Improved vigilance and market surveillance

Role & Competency of Notified Bodies

EU database -device traceability (UDI)

New risk classification system -IVDs

New requirments-clinical evidence

Introduction of an "implant card"

A seismic scandal with regard to fraudulent and adultered production of PIP silicone breast implants highlighted weaknesses in the legal system at the time and damaged the confidence of patients and healthcare professionals alike.

The introduction of the MDR rules and regulations should ensure such incidents are avoided in the future for products manufactured in the EU.

In addition, the EU aims to consolidate their rule as a global leader in the sector with a regulatory framework that keeps up to date with new technology and capabilities.

The new regulations will ensure:

o a consistent and high level of health and safety protection for EU citizens using medical devices/products

o free and fair trade of the products throughout the EU block

o EU legislation is updated and modernised to address the significant technological and scientific progress occurring in the medical device sector.

Implementation

From 26 November 2017, conformity assessment bodies (notified bodies) were illegible to apply for designation as notified bodies under Regulations (EU) 2017/745 and 2017/746.

Guidance documents and forms are provided by the *MDCG*. A joint assessment conducted onsite is also a requirement for designation of a recognised notified body.

A list of essential implementing acts/actions for the transitional period as well as information on expected timelines are detailed via a rolling plan detailing implantation measures which can be reviewed throughout the pre-launch and transition period for MDR at the EU commission internet website   location   -
https://ec.europa.eu/docsroom/documents/3404 1

Competent Authorities for Medical Devices project (CAMD)

An MDR/IVDR roadmap, produced by the Competent Authorities for Medical Devices project (aka CAMD) and the European Commission is also available for implementation guidance.

(CAMD recommended the establishment of an implementation task force)

CAMD Implementation Taskforce

The (CAMD) Executive Group recommended the establishment of an MDR-IVDR implementation taskforce to facilitate collaboration and cooperation within the medical devices industry network during the implementation stage of the new EU Regulations. As part of the implementation stage a consistent and harmonised approach in interpretation of the regulations across the network is fundamental to ensure consistency in regard to the application of the same.

Objectives

Establish a Regulatory system that is:

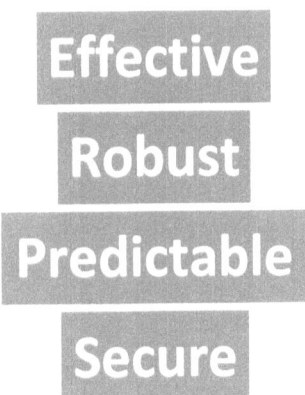

Effective

Robust

Predictable

Secure

An effective regulatory system is based upon the detail and depth of the legislation. Legislation is the basis for prosecuting medical device manufacturers that fail to comply with rules and regulations that are designed to ensure safe and effective devices. It is suggested that protection afforded to patients is the key measure of effectiveness. Robustness should allow a framework to respond to changes within the industry and technology, while a predicable framework fosters consistency in outcomes which benefit manufacturers and patients and end users.

(Official EU reference website)- https://www.camd-europe.eu/regulatory/medical-devices-regulation-vitro-diagnostics-regulation-mdr-ivdr-roadmap/

Medical Device Coordination Group

The Medical Device Coordination Group (MDCG) is an expert group established by Regulation, (EU) 2017/745 on medical-devices and Regulation (EU) 2017/746 on in- vitro diagnostic medical devices. Members are experts representing competent authorities all EU countries. The MDCG provides advice and expertise, assisting the Commission/ EU in implementation of both regulations.

MDCG Terms of Reference

Article 103 (1) of Regulation (EU)2017/7451 establishes the Medical Device Coordination Group

(MDCG). Per Article 103(9) of Regulation (EU) 2017/745 and Article 98 of Regulation (EU) 2017/7462, the MDCG group carries out the tasks conferred on it under both Regulations. Article 105 of Regulation (EU) 2017/745 and Article 99 of Regulation (EU) 2017/746 define general tasks of the MDCG. Specific tasks and roles of the MDCG are laid down in various provisions of the Regulations. The terms of reference document covers the following topics:

MDCG Terms of Reference	Tasks & Roles
	Membership
	Chairing
	Working group
	3rd Parties
	Stakeholders
	Operation
	Conflict of interest
	Rules & Procedures
	Professional secrecy & Classified information
	Transparency
	Travel & Subsistence expenses

(Official EU Reference website)
http://ec.europa.eu/transparency/regexpert/index.cfm?do=groupDetail.groupDetailDoc&id=37277&no=1

MDCG Rules of Procedure

Having regard to the standard rules of procedure of expert groups, the following Rules of Procedure have been adopted:

MDCG Rules of Procedure	
	Point 1 Operation of MDCG
	Point 2 Working Groups
	Point 3 Third Parties
	Point 4 Stakeholders
	Point 5 Meetings
	Point 6 Agenda
	Point 7 Documentation for the meeting
	Point 8 Positions and deliberations
	Point 9 Written Procedure
	Point 10 Meeing minutes
	Point 11 Attendance list
	Point 12 Support to the MDCG
	Point 13 Conflicts of interest
	Point 14 Correspondence
	Point 15 Transparency
	Point 16 Access to documents

For the manufacturer or authorised representative, the terms of reference and rules of procedure do not impact directly on the application of the new MDR rules. However, they are included here for general information and reference purposes and completeness.

The MDCG is currently made up of 11 distinct **working groups**:

1. Notified Bodies Oversight, (NBO)
2. Standards
3. Clinical Investigation and Evaluation, (CIE)
4. Post-Market Surveillance and Vigilance (PMSV)
5. Market Surveillance
6. Borderline and Classification (B&C)
7. New Technologies
8. Eudamed – see the register of Expert Groups under the code E01309
9. Unique Device Identification, (UDI)
10. International Matters
11. In vitro diagnostic medical devices, (IVD)

Notified Bodies Oversight (NBO) – Terms of reference

The role of NBO are to provide assistance to the MDCG on issues relating to notified bodies and conformity assessments. A key requirement for all NBOs is whether they can provide consistent, effective and harmonised application and implementation of Regulation (EU) 2017/745 on medical devices (MDR) and Regulation (EU) 2017/746 on in-vitro diagnostic medical devices (IVDR).

Notified bodies are based in a number of countries that serve the medical device industry with regard to certification and assessments.

The NBO also facilitates the sharing of experience and coordination of administrative practice among authorities responsible for notified bodies (the designating authorities), in accordance with Articles 48(1) MDR / 44(1) IVDR.

NBO prepares draft best-practice documents and model forms relating to the activities of designating authorities, notified bodies and their conformity assessment(s) activities, for endorsement by MDCG.

Standards – Terms of reference

The standards Working Group provides assistance to the MDCG on issues relating to standardisation in particular harmonised standards referred to in Article 8 of Regulation (EU) 2017/745 on medical devices (MDR) and Article 8 of Regulation (EU) 2017/746 on in-vitro diagnostic medical devices (IVDR).

Clinical Investigation and Evaluation (CIE) – Terms of reference

Clinical Investigation and Evaluation (CIE) provides assistance to the MDCG relating to clinical investigation and evaluation of medical devices in accordance with Regulation (EU) 2017/745 (MDR).

Post-Market Surveillance and Vigilance (PMSV) – Terms of reference

The PMSV group assists the MDCG on issues such as:

PMSV activities:

Preparation of technical guidance

Co-ordination

Provides a sharing forum

Reviews incidents

Post market clinical follow

Advice on post marketing surveillance

Market Surveillance – Terms of reference

Market Surveillance assists the MDCG and EU commission in coordinating market surveillance. The group deals with an analysis of:

Implementation of requirements set out in Annex I of the MDR / and IVDR

General obligations of economic operators laid down in Chapter II of the MDR / Chapter II of the IVDR

Obligations of economic operators and conformity assessment -products that do not require the involvement of notified bodies

Borderline & Classification (B & C) – Terms of reference

The Borderline & Classification Working Group provides assistance to the MDCG on issues relating to:

Qualification of a product as a medical device / an accessory for a medical device per EU MDR

Qualification of products without an intended medical purpose in accordance with Annex XVI MDR

classification of devices in accordance with Annex VIII MDR

The group prepares draft guidance on qualification and classification, for endorsement by the MDCG.

Under the consultation mechanism – the so-called "Helsinki Procedure". It also acts as a forum exchange of information with regards to qualification/classification of devices

New Technologies – Terms of reference

The New Technologies working group provides assistance to the MDCG on the following issues:

New and emerging technologies for medical devices

Guidance and common specifications in the field as referred to in Article 9 of MDR

Reviews electronic instructions for use with medical devices

Screening for identification of novel, emerging technologies for medical/clinical potential

Eudamed – ref the register of Expert Groups under the code E01309- Refer to E01309

Unique Device Identification (aka UDI) – Terms of reference

The group coordinates its activities with other MDCG working groups as appropriate.

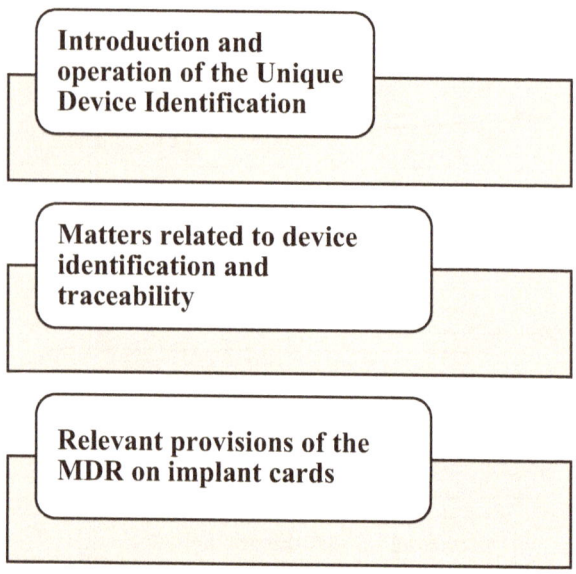

Introduction and operation of the Unique Device Identification

Matters related to device identification and traceability

Relevant provisions of the MDR on implant cards

International Matters – Terms of reference

International Matters Group provides the following assistance to the MDCG:

Monitors international regulation trends

Coordinates common views on harmonisation discussed within the International Medical Device Regulators Forum (IMDRF)

In vitro diagnostic medical devices (IVD) – Terms of reference

Assists MDCG on all IVD specific issues, e.g.homogenous application and implementation of Regulation (EU) 2017/746 (IVDR).

Prepares draft guidance on IVD related issues for endorsement of the MDCG

European Regulations on Medical Devices (MDR)

MDR Classification system shall see a number of new rules added to the existing requirements. There is also the introduction of a new IVDR classification system (A, B, C, D) A representing the least risk and D high risk. The new rules primarily relate to:

- o Software
- o Nanomaterials
- o Ingested products
- o Non-viable human tissues, cells & derivatives

The new Regulations place additional responsibilities and obligations on all parties in the sector.

Core purpose of the new regulation is the focus on increased availability of information on the identification, performance & safety of devices. A system called MDR -EUDAMED aims to facilitate a databank for medical devices.

Manufacturer

21

- Article 10
- Quality management
- system
- Risk management
- Clinical evaluation/PMCF
- Continued updates
- UDI & registration
- Labelling & language
- Incident reporting/FSCA
- Obligation to act
- Periodic reporting
- Liability cover for
- damage compensation

Authorised Representatives (AR)

- Article 11
- AR within the EU
- Written mandate – clear
- tasks
- AR legally accountable if
- the manufacturer fails to
- meet obligations
- Responsible Person

Person Responsible for Regulatory Compliance

- Article 15
- Demonstration of expertise
- Available within organisation
- Applies to Manufacturer and Authorised Representatives
- Responsible for checking conformity of devices

- o Updates to technical documentation
- o PMS obligations

- **Regulation (EU) 2017/745** of The European Parliament and of the Council of 5th April 2017 on medical devices, amending Directive 2001/83/EC, Regulation (EC) No 178/2002 and Regulation (EC) No 1223/2009 and repealing Council Directives 90/385/EEC and 93/42/EEC.
  (Reference- Official EU website)

https://eur-lex.europa.eu/legal-

content/EN/TXT/PDF/?uri=CELEX:32017R0745&from=EN

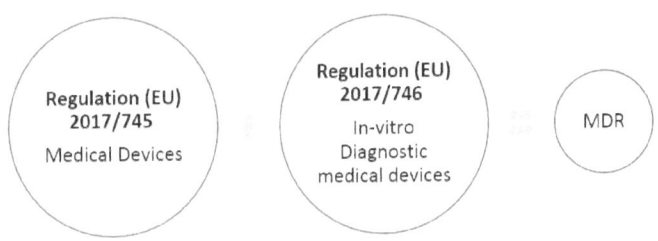

Components of the EU MDR

Regulation "No 1"

MDR-Medical Devices 2017/745

The new MDR Rule (collection of Chapters and articles and annexes I summarised below for ease of reference.

CHAPTER I -SCOPE AND DEFINITIONS

CHAPTER II -MAKING AVAILABLE ON THE MARKET AND PUTTING INTO SERVICE OF DEVICES, OBLIG

CHAPTER III -IDENTIFICATION AND TRACEABILITY OF DEVICES, REGISTRATION OF DEVICES AND OF ECONOMIC OPERATORS, SUMMARY OF SAFETY AND CLINICAL PERFORMANCE, EUROPEAN DATABASE ON MEDICAL DEVICES

CHAPTER IV- NOTIFIED BODIES

CHAPTER V-CLASSIFICATION AND CONFORMITY ASSESSMENT SECTION 1

Classification

SECTION 2-Conformity assessment

CHAPTER-VI CLINICAL EVALUATION AND CLINICAL INVESTIGATIONS

CHAPTER VII-POST-MARKET SURVEILLANCE, VIGILANCE AND MARKET SURVEILLANCE SECTION 1 Post-market surveillance

Article 83 Post-market surveillance system of the manufacturer

LABORATORIES, EXPERT PANELS AND
DEVICE REGISTERS

CHAPTER IX-CONFIDENTIALITY, DATA
PROTECTION, FUNDING AND PENALTIES

CHAPTER X-FINAL PROVISIONS

ANNEXES

ANNEX I GENERAL SAFETY AND
PERFORMANCE REQUIREMENTS CHAPTER I
GENERAL REQUIREMENTS

CHAPTER I DEFINITIONS SPECIFIC TO
CLASSIFICATION RULES
CHAPTER II IMPLEMENTING RULES
CHAPTER III CLASSIFICATION RULES

ANNEX IX CONFORMITY ASSESSMENT
BASED ON A QUALITY MANAGEMENT
SYSTEM AND ON ASSESSMENT OF
TECHNICAL DOCUMENTATION

ANNEX X CONFORMITY ASSESSMENT
BASED ON TYPE-EXAMINATION

ANNEX XI CONFORMITY ASSESSMENT
BASED ON PRODUCT CONFORMITY
VERIFICATION

ANNEX XII CERTIFICATES ISSUED BY A
NOTIFIED BODY

ANNEX XIII PROCEDURE FOR CUSTOM-
MADE DEVICES

ANNEX XIV CLINICAL EVALUATION AND
POST-MARKET CLINICAL FOLLOW-UP

ANNEX XV CLINICAL INVESTIGATIONS
ANNEX XVI LIST OF GROUPS OF PRODUCTS
WITHOUT AN INTENDED MEDICAL PURPOSE
REFERRED TO IN ARTICLE 1(2)

Regulation "No 2"

- **Regulation (EU) 2017/746** Of the European Parliament and of the Council of 5th April 2017 on in vitro diagnostic medical devices, and repealing Council Directives 98/79/EC Commission Decision 2010/227/EU.

(Official EU Reference)
https://eur-lex.europa.eu/legal-content/EN/TXT/PDF/?uri=CELEX:32017R0746&from=EN

CHAPTER I INTRODUCTORY PROVISIONS

Section 1
Scope and definitions
Article 1 Subject matter and scope
Article 2 Definitions

Section 2 Regulatory status of products and counselling
Article 3 Regulatory status of products
Article 4 Genetic information, counselling and informed consent
Article 5 Placing on the market and putting into service
Article 6 Distance sales
Article 7 Claims
Article 8 Use of harmonised standards
Article 9 Common specifications

CHAPTER III IDENTIFICATION AND TRACEABILITY OF DEVICES, REGISTRATION OF DEVICES AND OF ECONOMIC OPERATORS, SUMMARY OF SAFETY AND CLINICAL PERFORMANCE, EUROPEAN DATABASE ON MEDICAL DEVICES

CHAPTER V CLASSIFICATION AND
CONFORMITY ASSESSMENT

Section 1 Classification
Article 47 Classification of devices

Section 2 Conformity assessment
Article 48 Conformity assessment procedures
Article 49 Involvement of notified bodies in
conformity assessment procedures

Manufacturer Responsibilities

- Determine if your device is a medical device or a
  product without an intended medical purpose
  referred to in Annex XVI
- Determine the classification of your device per the
  requirements of Chapter V & Annex VIII, (Class I
  non-sterile, non-measuring and non- reusable
  surgical instrument medical devices and Class A in
  vitro medical devices do not require the intervention
  of a Notified Body)

- Identify the general safety and performance
  requirements that apply to your device (s) per Annex
  I

- Identify the harmonised standards and common
  specifications required to demonstrate compliance to

the general safety and performance requirements applicable to your device(s)

- Determine the conformity assessment route appropriate to your device(s) Annexes IX, X or XI as appropriate

- Determine the Technical documentation required to demonstrate compliance to the general safety and performance requirements applicable to your device(s)

- Develop your technical documentation in compliance to the above requirements per Annex II

- Review your post market surveillance, vigilance and market surveillance systems per the requirements of Chapter VII

- Develop your post market technical documentation per the requirements of Annex III

- Review your clinical evaluation and clinical investigations for compliance to Chapter VI 2017/745 for medical devices and clinical evidence, performance evaluation and performance studies for compliance to Chapter VI 2017/746 for in vitro diagnostic medical devices

- Review your clinical evaluation and post market surveillance for compliance to Annex XIV 2017/745 for medical devices and interventional clinical performance studies and certain other performance studies for compliance to Annex XIV 2017/746 for in vitro diagnostic medical devices

- Review your clinical investigations per the requirements of Annex XV 2017/745 for medical devices

- Draw up your declaration of conformity in compliance to Annex IV

- Develop your UDI –DI and UDI-PI

- Obtain an SRN

- Apply to a duly designated Notified Body under Regulation 2017/745 &/or 2017/746

Audit Implications for Manufactures

The regulation reinforces the authority and responsibilities of Notified bodies and how they interact with manufactures.

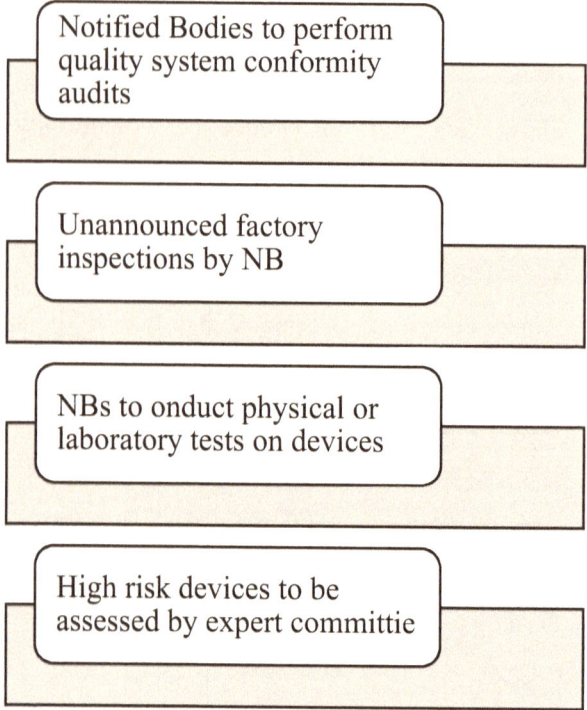

Notified Bodies to perform quality system conformity audits

Unannounced factory inspections by NB

NBs to onduct physical or laboratory tests on devices

High risk devices to be assessed by expert committie

Notified Body Responsibilities

With the introduction of MDR notified bodies will need to be fit for purpose and if they do not meet an adequate threshold, some notified bodies may not be re-certified. The first step requires them to apply for new designation.

It should also be highlighted that if notified bodies do not achieve certification, this will impact manufacturers of medical devices as they may need to switch to another notified body.

The requirement for notified bodies to complete unannounced audits of manufacturers and authorized representatives will increase, where audit schedules and details will have to be provided to national authorities ahead of time.

Notified Bodies will also be required to consult with the European Commission on the adequacy of the Notified Bodies clinical evaluation and post-market clinical follow-up plans, prior to the NBs granting certificates for Class III implantable and IIb devices intended to administer or remove medicines.

MDR EUAMED (European Databank on Medical Devices)

EUDAMED
European Databank on Medical Devices

| Electronic system on Registration | Electronic system on UDI | Electronic system on Certificates | Electronic system on Vigilance | Electronic system on Market surveillance | Electronic system on Clinical investigations |

The Commission will establish a centralised EU database (EUDAMED) for the storage of information on medical devices. This will facilitate the communication of both pre- and post-approval product information between:

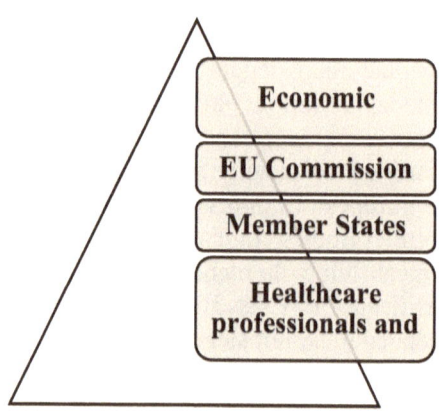

Economic

EU Commission

Member States

Healthcare professionals and

CE marking

What is CE Marking?

CE Marking is the abbreviation of French phrase "Conformité Européene" which means "European Conformity". Initially the term "EC Mark" was used and it but replaced by "CE Marking" in the Directive 93/68/EEC in 1993 with "CE Marking" now the standard lexicon.

Purpose

o CE Marking on a medical device is a manufacturer's declaration that the medical device complies with the essential requirements of the European health, safety and environmental legislation and directives.

o CE Marking placed on a medical device signifies to governmental officials that the device can be legally imported and placed on the EU market or within their country.

o CE Marking ensures free movement of the product within the EFTA, European Union (EU) single market and the EEA.

o CE Marking safeguards patients and users and legally permits withdrawal of non-conforming products by customs.

CE Marking and MDR Perspective

The regulation makes provision for manufacturers that any product issued with a valid CE Certificate, the certificate shall remain valid until the indicated expiry date, however, manufacturers are advised to transition sooner rather than later. Regulations shall come into force in May 2020. Notified bodies will be able to continue to issue certificates up until May 26, 2020 with such certificates having a validity until May 25, 2024.

Article 17, and affix the CE marking of conformity in accordance with Article 18.

Article 17
EU declaration of conformity

1. The EU declaration of conformity shall state that the requirements specified in this Regulation have been fulfilled. The manufacturer shall continuously update the EU declaration of conformity. The EU declaration of conformity shall, as a minimum, contain the information set out in Annex IV and shall be translated into an official Union language or languages required by the Member State(s) in which the device is made available.

2. Where, concerning aspects not covered by this Regulation, devices are subject to other Union legislation which also requires an EU declaration of conformity by the manufacturer that fulfilment of the requirements of that legislation has been demonstrated, a single EU declaration of conformity shall be drawn up in respect of all Union acts applicable to the device. The declaration shall contain all the information required for identification of the Union legislation to which the declaration relates.

3. By drawing up the EU declaration of conformity, the manufacturer shall assume responsibility for compliance with the requirements of this Regulation and all other European Union legislation applicable to the device.

4. The Commission is empowered to adopt delegated acts in accordance with Article 108 amending the minimum content of the EU declaration of conformity set out in Annex IV in the light of technical progress.

Article 18

CE marking of conformity

1. Devices, other than devices for performance studies are considered to be in conformity with the requirements of this Regulation shall bear the CE marking of conformity, as presented in Annex V to this MDR
2. The CE marking shall be subject to the general principles set out in Article 30 of Regulation (EC) No 765/2008.

3. The CE marking shall be affixed visibly, legibly and indelibly to the device or its sterile packaging. Where such affixing is not possible or not warranted on account of the nature of the device, the CE marking shall be affixed to the packaging. The CE marking shall also appear in any instructions for use and on any sales packaging.

4. The CE marking shall be affixed before the device is placed on the market. It may be followed by a pictogram or any other mark indicating a special risk or use.

5. Where applicable, the CE marking shall be followed by the identification number of the notified body responsible for the conformity assessment procedures set out in Article 48. The identification number shall also be indicated in any promotional material which mentions that a device fulfils the requirements for CE marking.

6. Where devices are subject to other Union legislation which also provides for the affixing of the CE marking, the CE marking shall indicate that the devices also fulfil the requirements of that other legislation.

Ref: https://eur-lex.europa.eu/legal-content/EN/TXT/?uri=CELEX%3A32017R0746

ANNEX IV

EU DECLARATION OF CONFORMITY

The EU declaration of conformity shall contain the following information:

1. Name, registered trade name or registered trade mark and, if already issued, SRN referred to in Article 28 of the manufacturer, and, if applicable, its authorised representative, and the address of their registered place of business where they can be contacted and their location be established

2. A statement that the EU declaration of conformity is issued under the sole responsibility of the manufacturer

3. The Basic UDI-DI as referred to in Part C of Annex VI

4. Product and trade name, product code, catalogue number or other unambiguous reference allowing identification and traceability of the device covered by the EU declaration of conformity such as a photograph, where appropriate, as well as its intended purpose. Except for product or trade name, information allowing identification and

traceability may be provided by the Basic UDI-DI referred to in point 3 above

5.Risk class of the device in accordance with the rules set out in Annex VIII;

6.A statement that the device that is covered by the present declaration is in conformity with this Regulation (EU MDR) and, if applicable, with any other relevant Union legislation that provides for the issuing of an EU declaration of conformity.

7.References to any CS used and in relation to which conformity is declared;

8.Where applicable the name and identification number of the notified body, a description of the conformity assessment procedure performed and identification of the certificate or certificates issued;

9. Where applicable, additional information;

10.Place and date of issue of the declaration, name and function of the person who signed it as well as an indication for, and on behalf of whom, that person signed, signature.

ANNEX V- CE MARKING OF CONFORMITY

1. The CE marking shall consist of the initials 'CE' taking the following form:

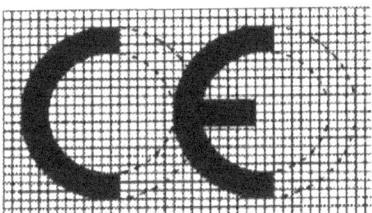

2. If the CE marking is reduced or enlarged the proportions given in the above graduated drawing shall be respected.

3. The various components of the CE marking shall have substantially the same vertical dimension, which may not be less than 5 mm. This minimum dimension may be waived for small-scale devices.

Medical Device Coordination Group

Medical Device Coordination Group was established under Article 103 of Regulation

('MDCG'). Their role is to determine, whether or not, a specific product, or category/ group of products, falls within the definitions of 'medical device' or as an 'accessory for a medical device'.

UDI (Unique Device Identification)

For the new MDR, all medical devices shall need to be fully traceable via the means of a Unique Device Identification (UDI) system. This has a profound implication for manufactures as means detailed planning for UDI implementation across the EU will be required. A similar system is operated in the United States. UDI system is detailed in Article 24 and with the registration obligations referred to in Article 26 and 28.

The traceability of each device will require a label on the device itself and not external packaging of shipping containers.

The UDI will be detailed as required in Vigilance Reports and thus will be used for reporting serious incidents /field safety actions. In similar fashion, implant cards with warnings and information about expected device lifetime, and follow-ups will be required for implantable medical devices.

Scope & Classification

The requirements with regard to Scope and Classification of Products is addressed in

Article 1 Subject matter and scope

Article 2 Definitions
Article 22 Systems and procedure packs
Article 23 Parts and components

Article 51 Classification of devices
Article 53 Involvement of notified bodies in
conformity assessment procedures

Annex' including VIII, IX, X, XVI.

Introduction of Strict Rules for substance-based devices and devices that use hazardous substances

Classification Rule 21, Annex I

Devices composed of substances/ combinations of substances intended to be introduced into the human body, via a body orifice or applied to the skin, and that are absorbed by or locally dispersed in the human body, will have a different classification depending on different factors.

Risk based classification

A new system for risk classification, in line with international guidelines and ISO standards, will apply to in vitro diagnostic medical devices.

Clinical Evaluations, Safety and Performance

Summary of Safety and Clinical Performance- (Article 32)

For implantable devices and for Class III devices, manufacturers must create a summary of safety and clinical performance. The summary must be written in a manner that is clear to the intended user (e.g. patient). The summary of a safety and clinical performance must include at least the following information:

- Manufacturer Name and Single Registration Number (SRN)
- Device Name and UDI
- Description, previous variant(s), differences
- Intended Purpose, Indications, contraindications and target population
- Possible diagnostic or therapeutic alternatives
- Reference to any harmonized standards and common specifications applied
- Summary of the clinical evaluation report (referred to in Annex IV) and relevant information on post-market clinical follow-up
- Suggested profile and training for users
- Information on residual risks, undesirable effects, warnings and precautions

Safety and Performance Requirements for Manufacturers

Annex I requirements on Safety and Performance replaced the Directive's Essential Requirements. Medical device manufacturers will have to demonstrate

compliance by showing documented evidence and practice of risk management, physical and chemical and development.

Clinical Evaluation / Post-market Clinical Follow-up and Clinical Investigations

Per Article 2, Article 55, Article 61, Annex XIV, Annex XV)

What is clinical evaluation?

Clinical evaluation for medical devices is a continuous procedure to collect, review, and analyse clinical data for a medical device. Clinical evaluation also includes testing to determine if relevant essential requirements for safety and performance are met when using the manufactures instructions for use.

What are the common sources used for clinical evaluation?

Sources for clinical evaluation may take account of published data, non-company clinical investigations, post-marketing surveillance data, adverse event databases (e.g., MAUDE) for equivalent devices or internal systems such as non-conformances or CAPA.

A Clinical Evaluation Report (CER) is a new requirement under MDR which gather safety and performance data.

Benefits of clinical evaluation

clinical evaluation is required for the purposes of CE marking, which is required to sell medical devices in EU member states and the wider European Economic Area.

It is understood that the CER is to be included in the CE technical file for CE marking/conformity assessment process.

Stages of clinical evaluation

Clinical evaluation consists of 5 stages:

Stage 0: Transition strategy and plan, gap analysis

Stage 1: Identification of data — which data is important to collect during development

Stage 2: Collection and analysis of clinical data — patient based

Stage 3: Establishing a CER process entailing all data

Stage 4: Periodic updates — the CER is updated over time and as required.

There are revised rules on clinical performance evaluation, and clinical investigations will require a thorough review of clinical strategy and post-market clinical follow-up plans. The current MEDDEVs on clinical requirements will be insufficient and will not comply with new rules.

Clinical evidence shall need to be updated for existing devices already on the market.

Class III devices and implantable devices- the post-market clinical follow-up evaluation report, the summary of safety and clinical performance shall be updated at least annually.

All requirements related to clinical data and clinical evaluations are defined in article 2 and article 61 and Annex XIV.

Notified bodies (NBs) will conduct reviews of the clinical investigation data for conformity to Annex XV per MDR.

Clinical Evaluations and MEDDEV

The 2016 MEDDEV 2.7/1 rev 4 was published with expanded requirements for clinical data, and the EU MDR reinforces this requirement. The EU MDR imposes much tighter restrictions on "device equivalency" for purposes of Clinical Evaluation Reports (CERs). You will want to review your CER very carefully to ensure compliance with the new requirements. There are additional requirements to report specific events to the upcoming European medical device electronic reporting system (EUDAMED).

Post Market Clinical Follow-up

Post-market clinical follow-up is a continuous process that updates the clinical evaluation and must be addressed as part of post-market surveillance plans, looking at data throughout the device's expected lifetime to help the manufacturer determine whether risks associated with the device remain acceptable and meets the intended purpose safely.

Usability

Usability or ergonomic "design" was listed within the Essential Requirements (Annex I) of the 1993 Medical Device Directive. However, with the new MDR, requirements for safe use and ease of use. Usability is often referred to as human factor testing or more generally as ergonomics.

Chapter I – General Requirements, Section 1

- ○ *"performance intended normal conditions of use, they are suitable for their intended purpose.*

- ○ *"safe and effective"*

- ○ *"shall not compromise the clinical condition"*

- ○ *"safety of patients, or the safety and health of users or, other persons, where applicable"*

- ○ *"Remaining risks constitute acceptable risks when weighed against the benefits"*

This requirement captures the intended use of the device and that it is fit for purpose. It also expands upon the impact of the medical device not only applies to the patient but to the "user" or other person".

This means that defining both "users" and "other persons" should be defined during the medical device development and design process and then evaluated during design control and post-launch to ensure the assumptions are accurate.

Annex I Chapter 1, (C)

- o *"Estimate/ evaluate the risks associated with and occurring during, the intended-use" and*

- o *"During reasonably foreseeable misuse"*

A usability risk assessment is required by the manufacturer to identify, assess and document use-based risks. This Risk assessment should be driven by ISO 14971:2012. If compliance to usability engineering standard IEC 62366-1:2015 is required then more scrutiny of root causes is necessary or use of PCA analysis (perceptive error cognitive analysis)

Annex I Chapter 1, Section 5(a).

In eliminating/reducing risks related to use- error, the manufacturer shall:

- *"reduce as far as possible the risks related to the ergonomic features of the deice and the environment in which the device is intended to be used" (design for patient safety)*

- *"give consideration to the technical knowledge, experience, education, training and use environment, where applicable, and the medical and physical conditions of intended users"*

  (designed for lay, professional, disabled or other users).

This requirement falls under the "human factors" and usability testing. It states that the manufacturer shall reduce the risk due to "ergonomic features" and to give consideration to human factors including:

- *"technical knowledge,*
- *experience,*
- *education,*
- *training,*
- *environment medical and physical conditions of the intended users"*
- *professional, lay, disabled or other users"*

Annex I Chapter 1, Section 6

Where "The characteristics and performance of a device shall:

"not be adversely affected to such a degree that the health or safety of the patient or user and where applicable, of other persons are compromised during the lifetime of the device as indicated by the manufacturer, when the device is subjected to the stresses which can occur during normal conditions of use and has been properly maintained in accordance with the manufacturer's instructions".

This requirement is a fundamental one which is often validated by completing a usability study. These studies examine the device over the lifetime of its intended use. Some studies may be completed in a simulated environment or with technology.

Annex I Chapter 2, Section 11.4.

Section 11.4 states that Devices delivered in a sterile state shall be

○ *"designed, manufactured and packaged in accordance with appropriate procedures*

○ *sterile when placed upon the market and that, unless the packaging which is intended to maintain their sterile condition is damaged, they remain sterile, under the transport and storage conditions specified by the manufacturer, until that packaging is opened at the point of use.*

o   *the integrity of that packaging is clearly evident to the final user."*

Here, the requirement examines the design manufacture and packaging of sterile medical devices, along with the integrity of the sterile packaging during transport and storage conditions. A component of usability testing is to examine the risk to sterility of the device.

Annex I Chapter 2, Section 14.2

The regulations states that

o   *"Devices shall be designed and manufactured in such a way as to remove or reduce as far as possible;*

o   *the risk of injury, in connection with their physical features, including the volume/pressure ratio, dimensional and where appropriate ergonomic features, and*

o   *the risks associated with the possible negative interaction between software and the IT environment within which it operates and interacts."*

The purpose of this requirement is to ensure that a medical device is developed where specific risks are minimised such as the environmental conditions, contact, software, interference, ingress and so on.

Annex I Chapter 2, Section 14.4

This requirement specifies that "Devices shall be designed /manufactured in a way that certain activities can be completed with ease such as:

- ○ *adjustment*
- ○ *calibration*
- ○ *maintenance*

Testing these aspects often fall under usability and human factors studies Section 14.5.

Annex I Chapter 2, Section 14.6

Section 14.6 relates to

- ○ *Measurement*
- ○ *Monitoring*
- ○ *Displaying*

By ensuring that ergonomic principles are involved in the design and manufacturing of the device.

Section 14.6 relates to:

- ○ *safe disposal*
- ○ *safe disposal of related waste substances by the user, patient or other person*

Therefore, it is the responsibility that manufacturers shall "identify and test" procedures and measures as a result of which their devices can be safely disposed after use and be listed in the instructions for use."

Annex I Chapter3, Section 1 Section 21.3:

Where the "*function of the controls and indicators shall be clearly specified on the devices.*

Where "*a device bears instructions required for its operation or indicates operating or adjustment parameters by means of a visual system, such information shall be understandable to the user and, as appropriate, the patient.*"

Controls and indicators must be clearly labelled and identified. Not only should say the medical professional be able to understand displays but also patients to a certain appropriate degree.

Any software using visual interaction or a HMI is required, usability testing can provide data and evidence that it is safe, usable and clear.

Part 2

Implementation Strategies

Introduction

As timelines move towards implementation, it is the concern of manufactures to developing suitably clear implementation strategies that will achieve compliance to MRD. It will require action and planning on many levels within a company to sufficiently prepare for the MDR regulations. Some key steps include:

- conducting a gap assessment of current system against new requirements.
- Assign resources that can begin to broaden their knowledge of the regulations. A cross functional team should meet regularly to discuss implementation.
- Develop a detailed written plan to document changes required and track their progress.

  It may be necessary to establish sub-groups to give adequate focus on MDR articles e.g. Product classification, Safety and performance, Post Marketing surveillance.

- classification
- Engage with NB and communicate and transition plan with them.

Risk Management

The risk management is a requirement of ISO 13485 and the essential requirements of the 93/42/EEC Medical Device Directive, the 98/79/EC IVD Directive. As previously outlined, 93/42/EEC Medical Devices and 98/79/EC IVD are to be repealed and replaced with Regulation (EU) 2017/745 and Regulation (EU) 2017/746 respectively.

Per Article 10 General obligations of manufacturers are as follows:

"Manufacturers shall establish, document, implement and maintain a system for risk management as described in Section 3 of Annex I."

Also, for Post-market surveillance system of the manufacturer

Article-89 Evaluation of devices suspected of presenting an unacceptable risk or other non-compliance Where the competent authorities of a Member State, based on data obtained by vigilance or market surveillance activities or on other information, have reason to believe that a device: (a) may present an unacceptable risk to the health or safety of patients, users or other persons, or to other aspects of the protection of public health; or (b) otherwise does not comply with the requirements laid down in this Regulation, they shall carry out an evaluation of the device concerned covering all requirements laid down in this Regulation relating to the risk presented by the device or to any other non-compliance of the device. The relevant economic operators shall cooperate with the competent authorities.

As previously described, a new risk-based classification system will apply to in-vitro diagnostic medical devices, in addition to a wider medical device classifications definition for all products.

While the classification system (Class III, Class IIa, Class IIb and Class I) will be retained, some rules have been tightened.

ISO 14971:2009 is the recognised standard for the risk management of medical devices.

Risk Analysis

Risk analysis can be performed using a variety of methodologies such as FMEA/FMECA, HAZOP, HACCP, or other methods appropriate to the design and function of the product. A common methodology for risk analysis for regulated products is FMEA. The analysis can be grouped into a product category of similar established device technology.

In estimating risk(s) for each hazard, information

may be obtained from the following sources:

- Clinical trial data
- Technical data
- Field data from similar products
- Usability tests
- Results of investigations, e.g. CAPA
- Expert opinion
- External assessments or audits

Severity (S): the severity score addresses how
                                        severe
the effect of this failure is on all/or one of the
following: the patient, operator, environment,
process, handlers and business. This is subjectively
rated from 1 to 5.

Occurrence (O): refers to how often the failure is
expected to occur. This can be rated either
subjectively (frequent, rare etc.) or via the number
of units affected depending on the risk being
assessed. This is rated from 1 to 5.

Detection (D): can the problem be detected by the
user or patient before it does any damage? This
column is subjectively rated from 1 to 4.

Calculate the Risk Priority Number (RPN) by
multiplying the numbers obtained for severity,
occurrence and detection together.

Risk Acceptability

RPN scores of risk acceptability can be divided into three categories:

Acceptable risk – risk is deemed acceptable meaning that there is no need to consider risk reduction measures.

Investigate further risk reduction – investigate if further risk reduction to the "no need to consider level" is practicable.

Unacceptable risk – risk control measures must be implemented to reduce risk.

Risk Control

Where risks are identified as unacceptable, risk control measures must be determined to reduce the risk prior to the process or system being implemented. A number of actions can be taken in order to further reduce risk including: (1) changing the design to reduce risk, introducing protective measures in the device or the manufacturing process, (3) inserting a warning statement into the instructions for use (IFU).

Risks scored as "investigate further risk reduction" should be examined to determine whether it is practicable to reduce the risk further. The risk should be reduced to as low as is reasonably practicable, (aka ALARP) taking into account the benefits of accepting the risk and the practicability of implementation. If risks classed as "investigate further risk reduction" are already at ALARP, no further risk reduction is necessary.

Residual Risk Evaluation
After risk control measures are applied, a new risk assessment will be carried out to determine residual risks. Residual risks will be assessed for acceptability using the same criteria as detailed in 6.5. If the residual risk is not judged acceptable then further risk control measures will be applied.

If the residual risk is not judged acceptable and further risk control is not practicable then the team may perform a risk/benefit analysis by evaluating data and literature on the medical benefits of the intended use to determine if they outweigh the risk. If this evidence does not support the conclusion that the medical benefits outweigh the residual risk, then the risk remains unacceptable. This analysis should be recorded and approved by both the risk management team and senior site management.

The Role of Data

A pathway to achieving compliance with EU MRD

Data: any data (numerical or otherwise) which is collected or processed as part of GxP activities in order to generate GxP documents and records using a paper-based or electronic process.

Data Handling: Any GxP task that involves creation, entry, review, approval, analysis, reporting, storage, archival, retrieval, or disposal of GxP data.

Data Life Cycle: Starts from the time of data creation to the point of use and during its retention, archival, retrieval and eventual disposal

GxP Impacting: Any action that can impact the quality or safety of a product or critical process.

Application: Software installed on a defined platform/hardware providing specific functionality.

Bespoke/Customised Computerised System: A computerised system individually designed to suit a specific business process.

Commercial Off-the-Shelf Software: Software commercially available, whose fitness for use is demonstrated by a broad spectrum of users.

IT Infrastructure: The hardware and software such as networking software and operation systems, which makes it possible for the application to function.

Life Cycle: All phases in the life of the system from initial requirements until retirement including design, specification, programming, testing, installation, operation and maintenance.

Data reliability is the foundation to achieving cGxP data integrity. The FDA's ALOCA model can be used to enforce data reliability.

Accuracy: the GxP data is recorded, calculated, analysed, and reported as found and correctly.

Attributable: any actions or calculations performed on GxP data can be attributed to or traceable to the person that performed the actions and the date and time at which they were performed.

Legible: the GxP data is recorded in a clear and human-readable form.

Contemporaneity: the GxP data is recorded at the same time as the observation/measurement is made or as soon as possible after the event.

Original: the initial data recorded is available and not altered.

Data Creation: The point at which the values or data is created. The data and information is original (raw).

Data Authentication: Within a GxP environment, authentication refers to the approval of data (electronic signatures). E-signatures are key controls within software that prompt the user to enter a unique username and password to acknowledge a recording or action. The e-signature should create a permanent link with the electronic record that cannot be removed and can be viewed through an audit trial.

Data Protection: Once the data is created, the handling of the data must ensure data integrity. For electronic data, this includes access control to computer systems. Other practical restrictions can also be made such as limiting room and site access to authorised personnel.

Data Retention: This refers to the controlled storage, backup and arching of data. Retention of records may be required for several decades depending on the type of data and the regulatory requirements relating to the particular product or industry.

Data Integrity

Data generated by or used in GxP impacting activities must be handled and protected in accordance with international and national regulatory requirements. The below agencies and regulatory authorities provide specific requirements on data integrity:

> ➤ EU GMP – EudraLex – Rules Governing Medicinal Products in the European Union Volume 4 – Guidelines to Good Manufacturing Practice for Medicinal Products for Human Use – Products for Human and Veterinary Use, Annex 11: Computerised Systems – (1, 7.2, 17)

> ➤ FDA – 21 CFR Part 11 – Food and Drug

Administration – Electronic Records; Electronic

Signatures – Scope and Application (C)

➢ FDA- 21 CFR Part 211 – Food and Drug Administration – Code of Federal Regulations -
Good Manufacturing Practices - 211.188a, 211.194.2, 211.194.8

➢ MHRA – United Kingdom - Medicines and Healthcare Products Regulatory Agency - GMP
Data Integrity Definitions and Guidance for Industry (2015)

The integrity of data relies on several factors. It can be influenced by a company's culture or approach to doing business. It can also be affected by the level of experience or knowledge within a company. Many traditional engineering companies outside the regulated life science community simply do not have the need to be so thorough in their handling of data and information.

Configuration Identification

Software and hardware packages should be identified by a unique product identifier and a version number. For the software end-user, the parts of an automated system that are subject to configuration management should be clearly identified. The system should therefore be broken down into configuration items. These should be identified at an early phase of development so that a complete list of configuration items is defined and maintained. The application-specific items should have a unique name or version ID. The depth of detail when specifying the elements is decided by the needs of the system, and the organisation developing that system.

Requirements for the User ID and Password

User ID: The user ID of a system should have a minimum length agreed with the customer and should be unique within the system.

Password: A password should always consist of a combination of numeric and alphanumeric characters. When setting up passwords, the number of characters and a period after which a password expires should be stipulated. The structure of the password is normally selected to suit the specific customer. The configuration is described in the security settings section of password policy. Criteria for the structure of a password are as follows:

- ➢ Minimum length of the password
- ➢ Use of numeric and alphanumeric characters
- ➢ Case sensitivity

Audit Trail

The audit trail is a control mechanism of a system that allows all data entered or modified to be traced back to the original data. A reliable and secure audit trail is particularly important in conjunction with the creation, change or deletion of GMP-relevant electronic records. In this case, the audit trail must archive and document all the changes or actions made along with the date and time. Typical contents of an audit trail must be recorded and describe the procedures "who changed what and when" (old value/new value).

The Life Cycle of Data

Regulations that speak to GxP and data integrity can apply to many different streams within the life science sector as previously mentioned. From medical devices to pharmaceuticals, all act in different manners, with long and short term applications. Take the example of a total knee replacement. Many designs now ensure their effectiveness in excess of ten years, even up to twenty years depending on individual circumstances. This requires many key records within manufacturing to be kept for several decades. Thus, data retention requirements specify the retention periods of such documents. The integrity of GxP data must be protected during the entire data life cycle, from creation of the data and records to the eventual destruction of data after the retention period is fulfilled. Data integrity equally applies to:

> ➢ Equipment
> ➢ Computerised systems
> ➢ Test records
> ➢ Inspection records
> ➢ Material certificates

Data integrity ensures that patient safety, product quality, and product supplies are generated by the product life cycle processes.

Process Design

Failure to maintain data integrity can occur throughout the life cycle of data; however, a thoughtful design of systems can prevent breaches in data and restrict the severity of any attempts to alter data. Therefore, design should aim to include controls and preventative measures. At a high level, this can be achieved by:

> ➢ Limiting access to GxP events and data
> ➢ Standard Operating Procedures (SOPs)
> ➢ Training
> ➢ System owners

Technical Controls

The benefits of modern software and computerised systems allow robust and complex data handling and calculations to be completed. With this modern capability that is becoming more powerful comes more responsibility with regard to the use of data.

The computerised systems used to generate, gather or interpret GxP data must fulfil several criteria. First and foremost, they must be fit for the intended use. The software and hardware must be validated and proven to be consistent and reliable. Some general considerations for the use of computerised systems include:

> Systems designed to foster integrity of GxP data
> User requirements specification detailing the intended use and required functionality
> An approved vendor with certification to ISO 9001 or other quality management standards
> Software should meet the requirements of regulations such as FDA 21 CFR Part 11
> Written procedures on how automated processes function

It should not be an easy process for personnel to alter or corrupt data when using computerised systems. GxP-impacting computer systems should have controls that prevent unauthorised access along with audit trail history.

Audit trail design and configuration capture key critical processes, events, settings and information. This enables any investigations of quality events impacting data integrity to be reviewed and analysed.

Practical Elements to Data Integrity

Facilities and systems must be configured in a way that encourages compliance with principles of data integrity. Examples include:

➢ Availability of clocks for recording times.
➢ Access points to allow swift reference to GxP records at locations where tasks are completed.
➢ Control of raw data.
➢ Control of approved documents.

Organisational Controls

Regulated companies such as medical device, pharmaceutical and biotechnology companies are required to operate under a quality management system. For medical devices, ISO 13485 serves as a quality management system. Likewise, the FDA Code of Federal Regulations 21 CFR Part 211 functions as a QMS for finished pharmaceuticals.

Organisational controls for Data Integrity can address:

- ➢ Assessment of GxP computerised systems
- ➢ Management of GxP computerised systems
- ➢ Electronic Records Implementation and handling
- ➢ Use of Electronic signatures
- ➢ Quality Risk Management

Operational Factors

Operational factors refer to process or manufacturing errors, deviations or non-compliance to established procedures that may impact data integrity.

GxP data handling activities should be designed to limit human intervention. As with human intervention there can be errors or omissions. Furthermore, it may call into question the reliability of the data. Mistake-proofing methodologies should be developed to avoid human error related breaches in data integrity. As with any system or technology, training is a fundamental step. Building upon training, exposure to GxP data systems and on-the-job training all play a part in delivering a system that is robust and meets regulatory requirements. It is important to remind ourselves that while regulations are the driving force to comply with data integrity, the ultimate goal is always the protection and safety of the patient or end user of the product, medicine or treatment.

Software Validation

Where there is the potential to affect product conformance to requirements or where software or IT systems provide support to aspects of quality management, validation is required. Most companies categorise software validations to account for the different applications of software and IT systems. For example, enterprise systems, such as the drawing package SolidWorks would be validated in a different manner to manufacturing systems that contain software (a.k.a. embedded software).

"Embedded" software is where the software is integrated into the manufacturing equipment. Embedded software is typically validated during the equipment qualification stage, process validation stage or test method validation. Enterprise software falls outside of equipment or process validation but does require validation if it impacts product quality or is used to make quality decisions. Standalone systems such as ERP (Enterprise Resource Planning) systems also require validation.

System Categorisation

21 CFR Part 11

This section specifically covers the regulatory requirements of part 11 of Title 21 of the Code of Federal Regulations; Electronic Records; Electronic Signatures (21 CFR Part 11). Part 11 of the FDA CFR is relevant to "records in electronic form that are created, modified, maintained, archived, retrieved, or transmitted under any records requirements set forth in agency regulations."

As of 2007, several sections of the regulation have been identified as excessive and the FDA announced in guidance that it will exercise enforcement discretion on some parts of 21 CFR part 11. This has been welcomed by some manufacturers but it has also caused a degree of confusion. The requirements relating to access controls are the most fundamental requirements and are routinely enforced. The "predicate rules" that required organisations to keep records in the first place are still in effect. If electronic records are illegible, inaccessible, or corrupted, manufacturers are still subject to those requirements.

If a regulated firm keeps "hard copies" of all required records, those paper documents can be considered the authoritative document for regulatory purposes. This then means that the computer system is not in scope for electronic records requirements, although subject to predicate rules which still require validation. If the "hard copy" is to be identified as the authoritative document, the "hard copy" must be a complete and accurate copy of the electronic source. The manufacturer must use the hard copy (rather than electronic versions stored in the system) of the records for regulated activities.

Definition of Records

The FDA has deemed the following records or signatures in electronic format subject to 21 CFR part 11:

Records that are required to be maintained under predicate rule requirements and that are maintained in electronic format in place of paper format. On the other hand, records (and any associated signatures) that are not required to be retained under predicate rules, but that are nonetheless maintained in electronic format, are not part 11 records. Records that are required to be maintained under predicate rules, that are maintained in electronic format in addition to paper format, and that are relied on to perform regulated activities.

Records submitted to FDA, under predicate rules (even if such records are not specifically identified in agency regulations) in electronic format (assuming the records have been identified in docket number 92S-0251 as the types of submissions the agency accepts in electronic format). However, a record that is not itself submitted, but is used containing nonbinding recommendations in generating a submission, is not a part 11 record unless it is otherwise required to be 205 maintained under a predicate rule and it is maintained in electronic format.

Electronic signatures that are intended to be the equivalent of handwritten signatures, initials, and other general signings required by predicate rules. Part 11 signatures include electronic signatures that are used, for example, to document the fact that certain events or actions occurred in accordance with the predicate rule (e.g. approved, reviewed, and verified).

The above definitions are taken from the FDA guidance document entitled "FDA Guidance for Industry: 21 CFR Part 11 - Electronic Records and Electronic Signatures: Scope and Application, August 2003." This document also provides recommendations on documenting key decisions that may be taken in relation to 21 CFR Part 11 applicability and compliance.

Requirements and Specifications

The need for compliance to 21 CFR depends on the type of technology and level of automation and computerisation involved in the manufacturing process or other actives that are GxP-impacting. Does the system store electronic records? Does the system require a login? Is there an audit trial? If a complex system is to be procured, the requirements need to be communicated to the manufacturer as part of a user requirement specification and/or software requirement specification.

General Guidance on Requirement Specifications

While the quality system regulation states that design input requirements must be documented, and that specified requirements must be verified, the regulation does not further clarify the distinction between the terms "requirement" and "specification." A requirement can be any need or expectation for a system or for its software. Requirements reflect the stated or implied needs of the customer, and may be market-based, contractual, or statutory, as well as an organisation's internal requirements.

There can be many different kinds of requirements (e.g., design, functional, implementation, interface, performance, or physical requirements). Software requirements are typically derived from the system requirements for those aspects of system functionality that have been allocated to software. Software requirements are typically stated in functional terms and are defined, refined, and updated as a development project progresses. Success in accurately and completely documenting software requirements is a crucial factor in successful validation of the resulting software. *Page 6 Guidance for Industry and FDA Staff General Principles of Software Validation A Specification* is defined as "a document that states requirements." (21 CFR 820.3(y)). It may refer to or include drawings, patterns, or other relevant documents and usually indicates the means and the criteria whereby conformity with the requirement can be checked.

There are many different kinds of written specifications, e.g., system requirements specification, software requirements specification, software design specification, software test specification, software integration specification, etc. All of these documents establish "specified requirements" and are design outputs for which various forms of verification are necessary.

Validation of Computerised Systems

85

The requirement for computerised systems to be compliant to 21 CFR part 11 needs to be identified early on in the project to ensure that the vendor or supplier of the systems or equipment can develop and build a system that meets the requirements of 21 CFR part 11. Computer system validation can be divided into three distinct phases: (1) planning, (2) design and development, (3) verification and (4) retirement.

Planning: This phase involves the planning of the validation effort required to implement the system and identification of key milestones and requirements. It requires supplier assessments, assessments of the regulatory and system risks, supplier development of a validation approach and the identification of deliverables that will be generated to support the implementation and operation of the system.

Design and Development: This phase consists of the design, development and configuration of the hardware and software required to meet the system requirements. In the case of custom software, design and developmental testing is important to ensure proper functionality prior to verification testing.

Verification: This phase confirms that requirements and specifications have been met. Testing is required to ensure the system operates as intended. Upon successful testing and verification, the system can be released for use. Once verification activities have begun, any changes to the system must be managed through change control. In case of successful completion of the verification activities (i.e. any deviation has been evaluated and addressed), the system is released for effective use. The operation phase supports the need to maintain compliance and fitness for intended use after the system is accepted and released for use.

Retirement: This phase consists of the planning, executing and summarising of the events required for system shutdown. It includes the appropriate handling of the supporting documents and the data contained within the system. While described here as a separate phase, a system's retirement can be handled as part of a new system implementation or as a separate project.

Best practice when it comes to computer system validation is to adopt a life cycle approach which requires the completion of activities in a systematic way from system conception to retirement. Life cycle activities could be scaled according to system impact on product quality, patient safety and data integrity, system complexity and novelty, supplier assessment and business risk.

Definitions

Computer System: A computer/automated system consisting of the hardware, software, and network components, together with the controlled functions (personnel, procedures, and equipment) and associated documentation.

Computer System Validation: A process that confirms by examination and provision of objective evidence that the computer system conforms to user needs and intended uses. Computer system validation is a process for achieving and maintaining compliance with GxP regulations and fitness for intended use by adoption of life cycle activities, deliverables, and controls.

GxP-Regulated Computer Systems: Computer systems determined to have a potential impact on product quality, patient safety and data integrity; these systems are required to comply with the relevant GxP regulations.

Data Integrity: The degree to which data is reliable and without error. Data must be accurate, attributable, contemporaneous, original, legible and available. A breach of data integrity occurs when any person manipulates or distorts data and submits the results of that data as valid.

Predicate Rules: A predicate rule is any FDA regulation that requires companies to maintain certain records and submit information to the agency as part of compliance.

To gain a better understanding of the validation of computerised systems, consult the following publication: "FDA's Guidance for Industry and FDA Staff General Principles of Software Validation." See also industry guidance such as the GAMP 5 guide issued by ISPE for a useful reference.

Electronic Records

When it comes to the regulated industries such as the medical device industry, every process and procedure must be documented. Documentation ensures that everyone is working in the same manner with the same procedures. However, documentation is more than just writing down procedures and processes. It is also concerned with how documents are controlled, how they are updated and how they are stored.

Electronic Document management systems

Electronic document management systems aka EDMS are now the norm and gold standard for most medium to large organisations. Many companies that provide medical device manufacturers with an EDMS that can be customised to match the business processes particular to an organisation. With configurable or customisable software, validation and proper verification is important to ensure the system operates as intended. There are also regulatory requirements that stipulate the expectations and requirements of such systems. For example, the application of electronic signatures and the presence of audit trials. FDA 21 CFR Part 11 details the requirements with regard to electronic records and electronic signatures. For medicinal products in Europe, GMP V4 Annex 11 specifies similar requirements.

Audit Trail

Title 21 CFR details predicate rule requirements relating to documentation of, for example, date time, or the sequencing of events, as well as any requirements for ensuring that changes to records do not obscure previous entries. Making the decision on whether to apply audit trails, or other appropriate measures, or on the need to comply with predicate rule requirements should involve a justified and documented risk assessment. Any risk assessment should determine the potential effect on product quality and safety and the integrity of the record.

Record Retention

With regard to the part 11 requirements for the protection of records to enable their accurate and ready retrieval throughout the records retention period (11.10 (c)), persons must also comply with all applicable predicate rule requirements for record retention and availability such as (211.180(c) general requirements. The decision to follow 21 CFR part 11 should be justified and documented as part of a risk assessment and based on the value of the records over time. The FDA does not object to archiving of required records in electronic format to non-electronic media such as paper, or to a standard electronic file format (examples of such formats include, but are not limited to, PDF, XML, or SGML). Persons must still comply with all predicate rule requirements, and the records themselves and any copies of the required records should preserve their content and meaning. As long as predicate rule requirements are fully satisfied and the content and meaning of the records are preserved and archived, you can delete the electronic version of the records. In addition, paper and electronic record and signature components can coexist as long as predicate rule requirements are met and the content and meaning of those records are preserved.

Part 3

Additional Guidance

- **Commission Recommendation 2013/172/EU** of 5 April 2013 on a common framework for a unique device identification system of medical devices in the Union
- **Commission Recommendation 2013/473/EU** of 24 September 2013 on the audits and assessments performed by notified bodies in the field of medical devices

MEDDEVs

What are MEDDEVs?

MEDDEVs form part of a set of guidelines relating to questions of application of EU Directives on medical devices (MDD/MDR). They are not legally binding.

MEDDEVs, been general guidance documents published by the EU are therefore open to a degree of interpretation and levels of applicability for manufacturers. Below is a list of MEDDEV (1-24) guidance in respect of EU MDR.

1. **MEDDEV 2.1/1** - Defines Medical Devices, Accessories and Manufacturer

2. **MEDDEV 2.12-2 rev 2** - Post market clinical follow up for medical devices

3. **MEDDEV 2.2/1 rev 1** - Addresses EMC requirements

4. **MEDDEV 2.2/3 rev 3** - Discusses the Use By date

5. **MEDDEV 2.2/4** - Conformity Assessment of IVF and ART Products

6. **MEDDEV 2.4/1 rev 9** - Classification of medical devices

7. **MEDDEV 2.5/3 rev 2** - When a quality-related inspection of a subcontractor is needed

8. **MEDDEV 2.1/2 rev 2** - Application of the Active Implantable Device Directive

9. **MEDDEV 2.1/3 rev 3** - Demarcation between MDD and Medicinal Products Directive

10. **MEDDEV 2.1/4** - Discusses demarcation between the EMC and PPE Directives

11. **MEDDEV 2.1/5** - Addresses medical devices with a measuring function

12. **MEDDEV 2.10/2 rev 1** - Designation and monitoring of Notified Bodies

13. **MEDDEV 2.12-1 rev 8** - Guidelines on a medical devices vigilance system

14. **MEDDEV 2.14/1 rev 2** - Borderline issues between the IVD and Medical Device Directives

15. **MEDDEV 2.14/2 rev 1** - Dealing with IVD products for research use only

16. **MEDDEV 2.14/3 rev 1** - Requirements for e-labelling of IVDs

17. **MEDDEV 2.15 rev 3** - Committees and organizations related to medical devices

18. **MEDDEV 2.5/10** - Guideline For Authorised Representatives

19. **MEDDEV 2.5/5 rev 3** - Clarifies translation procedures

20. **MEDDEV 2.5/6 rev 1** - Defines homogeneity of production batches

21. **MEDDEV 2.5/9 rev 1** - Medical devices containing natural rubber latex

22. **MEDDEV 2.7/1 rev 4** - Clinical Evaluation: Guide For Manufacturers And Notified Bodies

23. **MEDDEV 2.7/2** - Guidelines for Competent Authorities for making a validation/assessment of a clinical investigation application under directives 93/42/EEC and 90/385/EEC

24. **MEDDEV 2.7/4** - Guidelines on Clinical Investigation: A guide for manufacturers and notified bodies

MEDDEV 2.1/1 - Defines Medical Devices, Accessories and Manufacturer

This MEDDEV provides Definitions of "medical devices" "accessory" and "manufacturer".

MEDDEV 2.12-2 rev 2 - Post market clinical follow up for medical devices

The purpose of this document is to guide manufacturers and Notified Bodies on how to carry out Post-Market Clinical Follow-up (PMCF) studies in order to fulfil 1 Post-Market Surveillance (PMS) obligations according to

- ○ *Medical Devices Directive (93/42/EEC) and*

- ○ *Active Implantable Medical Devices Directive (90/385/EEC).*

Clinical evidence is a key element of the pre-market conformity assessment process to demonstrate conformity to Essential Requirements, however, manufacturers are cautioned to recognise the limitations to the clinical data available in the pre-market phase such as number of subjects, controlled setting and heterogeneity

MEDDEV 2.2/1 rev 1 - Addresses EMC requirements

This document provides guidelines relating to the application of council directive 90/385/EEC on active implantable medical devices the council directive 93/42/EEC on medical devices in respect of EMC requirements for electro magnetic devices. At a minimum the standard EN 60601-1-2 should be applied as harmonised standard for EMC aspects of the essential requirements.

MEDDEV 2.2/3 rev 3 - Discusses the Use By date

The purpose of this document is to prescribe information on the label and/or the information provided with the device on any time limitation on the safe use of the device.

MEDDEV 2.2/4 - Conformity Assessment of IVF and ART Products

This document provides guidelines on conformity assessment of "In Vitro Fertilisation" (IVF) and Assisted Reproduction Technologies (ART) products and devices.

the first reported live birth using by in-vitro fertilization was the birth of Louise Brown in 1978, a need for regulation has developed.

IVF/ART products may be qualified and regulated as medical devices or medicinal products, depending on the products "principal mode of action". Therefore, qualifications are made on a case by case basis.

MEDDEV 2.4/1 rev 9 - Classification of medical devices

This document contains guidance for the application of the *"classification rules for medical devices"* historically set out in Annex IX of Directive 93/42/EEC

- o *Practical Relevance Of Classification*
- o *How To Carry Out Classification*
- o *Explanations Of Individual Rules*

MEDDEV 2.5/3 rev 2 - When a quality related inspection of a subcontractor is needed

An inspection on the premises of a manufacturer's subcontractor may be require depending on the role of that sub-contractor.

Notified Bodies can assess whether the subcontractor has a substantial involvement with the design and/or production of the device, and b) if the subcontractor is undertaking the supply of a part, material or service, which may affect the compliance of the device with the essential requirements.

MEDDEV 2.1/2 rev 2 - Application of the Active Implantable Device Directive

This document provides definitions of Active implantable medical devices and examples including common Accessories.
An active medical device is defined as "implantable" if it is "totally or partly introduced, either surgically or medically, into the human body or by medical intervention into a natural orifice, and which is intended to remain after the procedure".

MEDDEV 2.1/3 rev 3 - Demarcation between MDD and Medicinal Products Directive

This document provides guidance on Medical devices versus medicinal product. It also contains guidance on borderline classification issues.

MEDDEV 2.1/4 - Discusses demarcation between the EMC and PPE Directives

Provides Guidance on Electromagnetic compatibility in relation to Personal protective equipment.

MEDDEV 2.1/5 - Addresses medical devices with a measuring function

Medical devices with a measuring function or feature must fulfil specific criteria

- o *The device is intended by the manufacturer to measure - quantitatively a physiological or anatomical parameter,*
- o *quantity or a qualifiable characteristic of energy or of substances delivered to or removed from the human body.*

The result of measurement - is displayed in legal or other acceptable units

MEDDEV 2.10/2 rev 1 - Designation and monitoring of Notified Bodies

This document provides a framework for the Designation and monitoring of Notified Bodies. Again, updates to this guidance in likely in respect of MDR. The guidance covers:

o   Independence requirements
o   Impartiality requirements
o   Competence requirements
o   Internal   procedures   and   facilities   -
    requirements
o   Confidentiality requirements
o   Liability insurance
o   Subcontracts
o   Notified Body's quality system

MEDDEV 2.12-1 rev 8 - Guidelines on a medical devices vigilance system

This document sets out the requirements in respect if device vigilance. The guidelines are often updated with regulatory developments and other inputs for organisations such as the GHTF on medical devices.

MEDDEV 2.14/1 rev 2 - Borderline issues between the IVD and Medical Device Directives

The present guideline gives a non-exhaustive list of IVD medical devices, and accessories to IVD medical devices. The document provides General principles of qualification along with more Specific qualifications including:

o   *Microbiological culture media*
o   *Stains*
o   *Devices with an invasive body contact for IVD purposes*
o   *Devices where no specimen is involved*

o *Devices involved in biological or chemical warfare*
   o *Devices to be used in law enforcement*
   o *Relation with the Biocides Directive 98/8/EC*

Classification information is also detailed within the document.

MEDDEV 2.14/2 rev 1 - Dealing with IVD products for research use only (RUO)

This document provides guidance on Research-Use Only IVD medical products.

Use of "RUO" Labelled Products

(a) RUO products used initial/ Basic Research: Products used for research conducted to study all aspects of human life in an attempt to better understand all underlying mechanisms.

   In such studies OR experiments animal and / or human models are used

(b) RUO products used in Pharmaceutical Research: The products are used to verify the reactions to compounds in animal and / or human models. In such practice there is no potential to misuse RUO products.

(c) RUO products used for a better identification and quantification of individual chemical substances or ligands in biological specimens: These products usually react with substances in a specimen through specific bindings or chemical reactions e.g. monoclonal or polyclonal antibodies. The RUO products are not sold by the manufacturers with an intended use within the definition of an IVD.

(d) In-house manufacturing of so called "home brew kits". This may involve the use of laboratory tools such as primers to improve the performance of an existing IVD within a healthcare institution. The IVD Directive does not cover this type of research.

MEDDEV 2.14/3 rev 1 - Requirements for e-labelling of IVDs

This document deals with supply of Instructions For Use (IFU) and other information for In-vitro Diagnostic (IVD) Medical Devices, in particular fo when information to be supplied by Different Means of Supply.

MEDDEV 2.5/10 - Guideline For Authorised Representatives

The purpose of this guideline is to set out what the Directives currently says on the role and the responsibilities of authorised representatives and also to set out the Member States' expectations as to the role of the authorised representatives in terms of market surveillance.

MEDDEV 2.7/1 rev 4 - Clinical Evaluation: Guide For Manufacturers And Notified Bodies

The 2016 MEDDEV 2.7/1 rev 4.0 was published with expanded requirements for clinical data, and the EU MDR reinforces this requirement. Manufacturers of medical devices must plan, execute and document a clinical evaluation to demonstrate conformity with safety and performance requirements found in Annex 1. The clinical evaluation, and the related documentation, must be updated with post-market clinical data throughout the life cycle of the device.

MEDDEV 2.7/2 - Guidelines for Competent Authorities for making a validation/assessment of a clinical investigation application under directives 93/42/EEC and 90/385/EEC

MEDDEV 2.7/4 - Guidelines on Clinical Investigation: A guide for manufacturers and notified bodies

APPENDIX 1

Regulation (EU) 2017/745 -Listed summary of Chapters and articles

Regulation (EU) 2017/745 of The European Parliament and of the Council of 5th April 2017 on medical devices, amending Directive 2001/83/EC, Regulation (EC) No 178/2002 and Regulation (EC) No 1223/2009 and repealing Council Directives 90/385/EEC and 93/42/EEC.

CHAPTER I -SCOPE AND DEFINITIONS

Article 1 Subject matter and scope
Article 2 Definitions
Article 3 Amendment of certain definitions
Article 4 Regulatory status of products

CHAPTER II -MAKING AVAILABLE ON THE MARKET AND PUTTING INTO SERVICE OF DEVICES, OBLIG

Article 5 Placing on the market and putting into service
Article 6 Distance sales
Article 7 Claims
Article 8 Use of harmonised standards
Article 9 Common specifications
Article 10 General obligations of manufacturers
Article 11 Authorised representative
Article 12 Change of authorised representative
Article 13 General obligations of importers
Article 14 General obligations of distributors

CHAPTER III -IDENTIFICATION AND TRACEABILITY OF DEVICES, REGISTRATION OF DEVICES AND OF ECONOMIC OPERATORS, SUMMARY OF SAFETY AND CLINICAL PERFORMANCE, EUROPEAN DATABASE ON MEDICAL DEVICES

CHAPTER IV- NOTIFIED BODIES

CHAPTER V-CLASSIFICATION AND CONFORMITY ASSESSMENT SECTION 1

Classification

Article 51 Classification of devices

SECTION 2-Conformity assessment
Article 52 Conformity assessment procedures
Article 53 Involvement of notified bodies in conformity assessment procedures
Article 54 Clinical evaluation consultation procedure for certain class III and class IIb devices
Article 55 Mechanism for scrutiny of conformity assessments of certain class III and class IIb devices
Article 56 Certificates of conformity
Article 57 Electronic system on notified bodies and on certificates of conformity
Article 58 Voluntary change of notified body
Article 59 Derogation from the conformity assessment procedures
Article 60 Certificate of free sale

CHAPTER-VI CLINICAL EVALUATION AND CLINICAL INVESTIGATIONS

Article 61 Clinical evaluation
Article 62 General requirements regarding clinical investigations conducted to demonstrate conformity of devices
Article 63 Informed consent
Article 64 Clinical investigations on incapacitated subjects
Article 65 Clinical investigations on minors

CHAPTER VII-POST-MARKET SURVEILLANCE, VIGILANCE AND MARKET

CHAPTER VIII-COOPERATION BETWEEN MEMBER STATES, MEDICAL DEVICE COORDINATION GROUP EXPERT LABORATORIES, EXPERT PANELS AND DEVICE REGISTERS

CHAPTER IX-CONFIDENTIALITY, DATA PROTECTION, FUNDING AND PENALTIES

ANNEXES

ANNEX I GENERAL SAFETY AND
PERFORMANCE REQUIREMENTS CHAPTER I
GENERAL REQUIREMENTS

CHAPTER II REQUIREMENTS REGARDING
DESIGN AND MANUFACTURE
CHAPTER III REQUIREMENTS REGARDING
THE INFORMATION SUPPLIED WITH THE
DEVICE

ANNEX II TECHNICAL DOCUMENTATION

ANNEX III TECHNICAL DOCUMENTATION
ON POST-MARKET SURVEILLANCE

ANNEX IV EU DECLARATION OF
CONFORMITY

ANNEX V CE MARKING OF CONFORMITY

ANNEX VI INFORMATION TO BE SUBMITTED
UPON THE REGISTRATION OF DEVICES AND
ECONOMIC OPERATORS IN ACCORDANCE
WITH ARTICLES 29(4) AND 31, CORE DATA
ELEMENTS TO BE PROVIDED TO THE UDI
DATABASE TOGETHER WITH THE UDI-DI IN
ACCORDANCE WITH ARTICLES 28 AND 29,
AND THE UDI SYSTEM

ANNEX VII REQUIREMENTS TO BE MET BY
NOTIFIED BODIES

ANNEX VIII CLASSIFICATION RULES
CHAPTER I DEFINITIONS SPECIFIC TO
CLASSIFICATION RULES
CHAPTER I DEFINITIONS SPECIFIC TO
CLASSIFICATION RULES
CHAPTER II IMPLEMENTING RULES
CHAPTER III CLASSIFICATION RULES

ANNEX IX CONFORMITY ASSESSMENT
BASED ON A QUALITY MANAGEMENT
SYSTEM AND ON ASSESSMENT OF
TECHNICAL DOCUMENTATION

ANNEX X CONFORMITY ASSESSMENT
BASED ON TYPE-EXAMINATION

ANNEX XI CONFORMITY ASSESSMENT
BASED ON PRODUCT CONFORMITY
VERIFICATION

ANNEX XII CERTIFICATES ISSUED BY A
NOTIFIED BODY

ANNEX XIII PROCEDURE FOR CUSTOM-
MADE DEVICES

ANNEX XIV CLINICAL EVALUATION AND
POST-MARKET CLINICAL FOLLOW-UP

ANNEX XV CLINICAL INVESTIGATIONS

ANNEX XVI LIST OF GROUPS OF PRODUCTS
WITHOUT AN INTENDED MEDICAL PURPOSE
REFERRED TO IN ARTICLE 1(2)

APPENDIX 2

Regulation (EU) 2017/746

CHAPTER I INTRODUCTORY PROVISIONS

CHAPTER III IDENTIFICATION AND TRACEABILITY OF DEVICES, REGISTRATION OF DEVICES AND OF ECONOMIC OPERATORS, SUMMARY OF SAFETY AND CLINICAL PERFORMANCE, EUROPEAN DATABASE ON MEDICAL DEVICES

CHAPTER IV NOTIFIED BODIES

CHAPTER V CLASSIFICATION AND CONFORMITY ASSESSMENT

Article 49 Involvement of notified bodies in conformity assessment procedures

Glossary

A

Audit Trail

The audit trail is a control mechanism of a system that allows all data entered or modified to be traced back to the original data. A reliable and secure audit trail is particularly important in conjunction with the creation, change or deletion of GMP-relevant electronic records.

Acceptable Quality Level (AQL)

The AQL of a sampling plan is the Process Performance Level routinely accepted by the sampling plan.

B

Biocompatibility

A measure of how a biomaterial interacts in the body with the surrounding cells, tissues and other factors.

Bioburden

The level and type of micro-organisms that can be present in raw materials, API starting materials, intermediates or APIs. Bioburden should not be considered contamination unless the levels have been exceeded or defined objectionable organisms have been detected.

Biological Indicators

A test system containing viable microorganisms providing a defined resistance to a specified sterilisation process, e.g. vaporised hydrogen peroxide.

Borderline Classifications

In certain circumstances, it may not be clear if a product falls under the medical device legislation or whether to classify a device as a medicine, cosmetic, biocide and so on. The decision will largely depend on the particular intended use of the product, as assigned by the manufacturer, and on the demonstrated mode of action. The manufacturer's claims must be substantiated by relevant data.

BSI

British Standards Institute.

C

Calibration

A requirement that demonstrates a particular instrument or device produces results within specified limits by comparison with those produced by a reference or traceable standard over an appropriate range of measurements.

Change Control

A formal system by which qualified representatives of appropriate disciplines review proposed or actual changes that may impact the validated status.

Change Management

An overarching approach to change control that is used during the preliminary planning and design stage of a project.

Competent Authority

A competent authority is the legally designated authority mandated to monitor compliance with directives and legal requirements within the industry. The competent authority has the power to grant and revoke licenses.

Compendial Organisations

Organisations certifying material standards that meet compendial requirements and acceptance criteria, e.g. the United States Pharmacopeia.

CCC (Mark)

The China Compulsory Certificate mark, commonly known as a CCC Mark, is a safety mark for many products sold on the Chinese market. As of 2013, medical devices do not require this certification.

CE Marking

CE Marking is a mandatory conformance mark on many products (including medical devices) placed on the single market in the European Economic Area. The CE marking certifies that a product has met EU consumer safety, health or environmental requirements. By affixing the CE marking to a product, the manufacturer declares that it meets EU safety, health and environmental requirements.

CEN

Communité Européenne des Normes (European Committee for Standardisation).

Clinical Trial

Clinical trials are conducted to allow safety and efficacy data to be collected for health interventions (e.g. drugs, diagnostics, devices, therapy protocols). These trials can only take place after satisfactory information has been gathered on the quality of the non-clinical safety, and health authority/ethics committee approval is granted in the country where the trial is taking place.

Conformity

Fulfilment of a requirement or meeting a requirement.

Conformity Assessment Body (CAB)

A body, other than a regulatory (competent) authority, engaged in determining whether the relevant requirements in technical regulations or standards are fulfilled.

CRO

A "contract research organisation", also commonly known as a "clinical research organisation", is a service organisation that provides support to the pharmaceutical and biotechnology industries. CROs offer clients a wide range of "outsourced" pharmaceutical research services to aid in the drug and medical device research and development process.

D

Data Integrity

Refers to the degree to which data is reliable and without error. Data must be accurate, attributable, contemporaneous, original, legible and available. A breach of data integrity occurs when any person manipulates or distorts data and submits the results of that data as valid.

Design Controls

Design controls are a collection of practices and procedures that are incorporated into the design and development process for a product such as a medical device. They provide a structure and clear path from the user needs assessment to product delivery through a step-by-step process. Design controls ensure proper assessment of the design is completed during the design and development phase. Design controls are a requirement of quality systems such as 21 CFR Part 820 (medical devices), and for certain classes of devices and per ISO 13485 - Quality Management Systems.

F

H

Harm

Damage to health, including the damage that can occur from loss of product quality or availability.

ISO 14971

An ISO standard, published in 2007, that provides a framework and requirements for a risk management system for medical devices. This standard establishes the requirements for risk management to determine the safety of a medical device by the manufacturer during the product life cycle.

ISO 9001

ISO 9001 is an ISO standard that represents the requirements for quality management systems. It is used across industries and is not specific to medical devices like ISO 13485.

IVD

In vitro diagnostic tests are medical devices intended to perform diagnoses from assays in a test tube, or more generally in a controlled environment outside a living organism.

MDD

The Medical Device Directive is intended to harmonise the laws relating to medical devices within the European Union. Medical Device Directive 93/42/EEC was most recently reviewed and amended by 2007/47/EC.

MHRA

The Medicines and Healthcare Products Regulatory Agency (MHRA) is the UK government agency which is responsible for ensuring that medicines and medical devices work and are acceptably safe.

Notified Bodies

A notified body is a certification organisation which the national authority (the competent authority) of a member state designates to carry out one or more of the conformity assessment procedures or audits described in the annexes of the medical devices directives or GMP legislation.

PMS

Post marketing surveillance is the practice of monitoring a pharmaceutical drug or device after it has been released on the market.

Q

Quality Management System

A Quality Management System, often abbreviated to QMS, is any system based on a collection of business processes that are primarily focused on providing safe and quality products that consistently meet customer requirements.

U

UDI, Unique Device Identification

The UDI is a series of numeric or alphanumeric characters that is created through a globally accepted device identification and coding standard. It allows the unambiguous identification of a specific medical device on the market.

Additional References

Usability engineering standard IEC 62366-1:2015

ISO 14971:2012 Risk Management Medical Devices

Council Directive 93/42/EEC of 14 June 1993 concerning medical devices as last 9amended by Directive 2007/47/EC of the European Parliament and of the Council of 5 September 2007.

Council Directive 90/385/EEC of 20 June 1990 on the approximation of the laws of 95 the Member States relating to active implantable medical devices last amended by Directive 2007/47/EC of the European Parliament and of the Council of 5 September 2007

www.ingramcontent.com/pod-product-compliance
Lightning Source LLC
Chambersburg PA
CBHW021828170526
45157CB00007B/2720